# Disclaimer

The publisher of this book is by no way associated with the National Institute of Standards and Technology (NIST). The NIST did not publish this book. It was published by 50 page publications under the public domain license.

50 Page Publications.

**Book Title:** Analysis of Three Different Regression Models to Estimate the Ballistic Performance of New and Environmentally Conditioned Body Armor

**Book Author:** Diane Mauchant; Michael A. Riley; Kirk D. Rice; Amanda L. Forster; Dennis D. Leber; Daniel V. Samarov

**Book Abstract:** The performance standard for ballistic-resistant body armor published by the National Institute of Justice (NIJ), NIJ Standard 0101.06, recommends estimating the perforation performance of body armor by performing a statistical analysis on V50 ballistic limit testing data. The logistic regression model is the statistical regression model recommended in the NIJ standard, but depending on the armor system and the amount of ballistic data collected for that system, other regression models may be more appropriate. Thus, the first objective of this study is to evaluate and compare the estimations of the performance provided by different statistical methods applied to ballistic data generated in the laboratory. Three different distribution models that are able to describe the relationship between the projectile velocity and the probability of perforation are considered: the logistic, the probit and the complementary log-log response models. In this work, each regression model will be discussed and applied to the ballistic limit data, using the method of maximum likelihood to estimate regression parameters. Then the model estimation results will be compared and evaluated. Different criteria for assessing the goodness of fit for each model will be investigated to identify criteria that can best distinguish which regression method produces the most accurate estimate of the performance of a particular armor model and to understand how an armor model s performance changes as it ages, either through field or laboratory aging. A secondary objective of this study is to apply the different methods to a new body armor model with unusual ballistic limit results, leading one to suspect that it may not be best described by a symmetric model, to determine if this data can be better fitted by a model other than the logistic model.

**Citation:** NIST Interagency/Internal Report (NISTIR) - 7760

**Keyword:** body armor, statistics, logit, probit, c-log-log, ballistic limit, binary response data

NISTIR 7760

# Analysis of Three Different Regression Models to Estimate the Ballistic Performance of New and Environmentally Conditioned Body Armor

Diane Mauchant, Kirk D. Rice, Michael A. Riley, Dennis Leber,
Daniel Samarov, and Amanda L. Forster

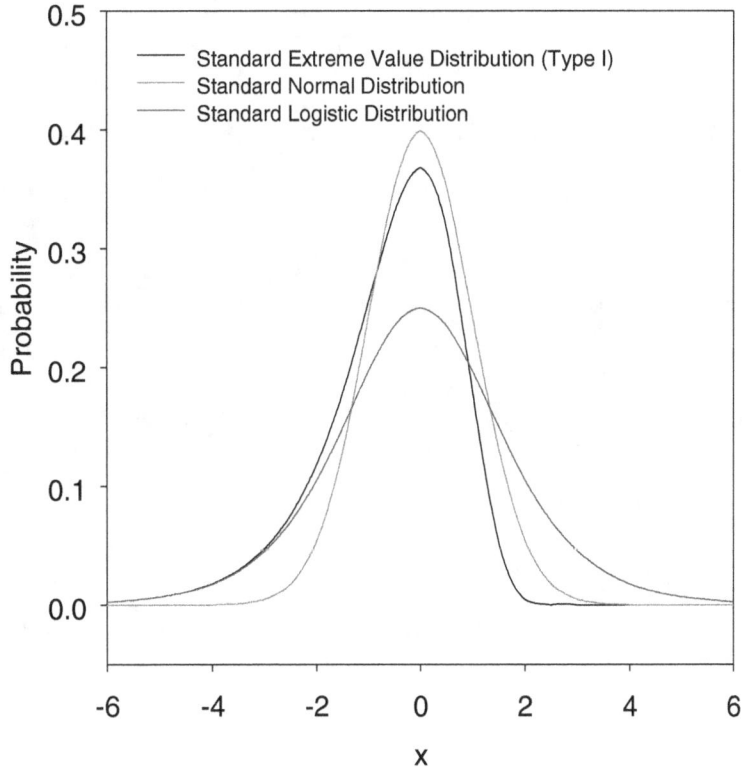

National Institute of
Standards and Technology
U.S. Department of Commerce

NISTIR 7760

# Analysis of Three Different Regression Models to Estimate the Ballistic Performance of New and Environmentally Conditioned Body Armor

Diane Mauchant, Michael A. Riley,
Kirk D. Rice, and Amanda L. Forster
*Law Enforcement Standards Office*

*Special Programs Office*

Dennis Leber and Daniel Samarov
*Statistical Engineering Division*

*Information Technology Laboratory*

February 2011

U.S. Department of Commerce
*Gary Locke, Secretary*

National Institute of Standards and Technology
*Patrick D. Gallagher, Director*

# Abstract

The performance standard for ballistic-resistant body armor published by the National Institute of Justice (NIJ), NIJ Standard–0101.06, recommends estimating the perforation performance of body armor by performing a statistical analysis on $V_{50}$ ballistic limit testing data. The logistic regression model is the statistical regression model recommended in the NIJ standard, but depending on the armor system and the amount of ballistic data collected for that system, other regression models may be more appropriate. Thus, the first objective of this study is to evaluate and compare the estimates of the performance provided by different statistical methods applied to ballistic data generated in the laboratory. Three different distribution models that are able to describe the relationship between the projectile velocity and the probability of perforation are considered: the logistic, the probit and the complementary log-log response models. In this work, each regression model will be discussed and applied to the ballistic limit data, using the method of maximum likelihood to estimate regression parameters. Then the model estimation results will be compared and evaluated. Different criteria for assessing the goodness-of-fit for each model will be investigated to identify criteria that can best distinguish which regression method produces the most accurate estimate of the performance of a particular armor model and to understand how an armor model's performance changes as it ages, either through field or laboratory aging. A secondary objective of this study is to apply the different methods to a new body armor model with unusual ballistic limit results, leading one to suspect that it may not be best described by a symmetric model, to determine if these data can be better fitted by a model other than the logistic model.

This page intentionally left blank.

# Acknowledgments

Financial support for this research effort was provided by the National Institute of Justice under Interagency Agreement Numbers 2003-IJ-R-029 and 2008-DN-R-121, by the Community Oriented Policing Services Office, and by NIST. Their support is gratefully acknowledged.

This page intentionally left blank.

# Disclaimer

Certain commercial equipment, instruments, or materials are identified in this paper in order to specify the experimental procedure adequately. Such identification is not intended to imply recommendation or endorsement by the National Institute of Standards and Technology, nor is it intended to imply that the materials or equipment identified are necessarily the best available for this purpose.

This page intentionally left blank.

# Contents

1 **Introduction**   1

2 $V_{50}$ **Ballistic Limit Data Analysis**   3
   2.1 Experimental . . . . . . . . . . . . . . . . . . . . . . . . 3
   2.2 The Regression Models . . . . . . . . . . . . . . . . . . . 4
      2.2.1 Background . . . . . . . . . . . . . . . . . . . . . . 4
      2.2.2 Presentation of the Three GLMs . . . . . . . . . . 5
   2.3 Application of the Different Models to Fit $V_{50}$ . . . . . . . . 7
      2.3.1 R Software . . . . . . . . . . . . . . . . . . . . . . 7
      2.3.2 Comparison of the GLM Estimates . . . . . . . . . 8

3 **Generalized Linear Model Estimation Evaluation**   13
   3.1 Assessment Criteria . . . . . . . . . . . . . . . . . . . . . 13
      3.1.1 Akaike's Information Criterion . . . . . . . . . . . 13
      3.1.2 Log-Likelihood . . . . . . . . . . . . . . . . . . . . 14
      3.1.3 Deviance of the Model . . . . . . . . . . . . . . . . 14
      3.1.4 Prediction Error Rate . . . . . . . . . . . . . . . . 14
      3.1.5 Cross-Validation Method . . . . . . . . . . . . . . 14
   3.2 Model Diagnostics Results . . . . . . . . . . . . . . . . . 15
   3.3 Results of Theoretical Analysis . . . . . . . . . . . . . . . 16

4 **Application to New and Conditioned Armor Analysis**   17
   4.1 Armor Comparison . . . . . . . . . . . . . . . . . . . . . 17
   4.2 Model Selection . . . . . . . . . . . . . . . . . . . . . . . 19
   4.3 Summary . . . . . . . . . . . . . . . . . . . . . . . . . . 20

5 **Application of the GLMs to an Unusual Data Set**   21
   5.1 Global Analysis of the Armor . . . . . . . . . . . . . . . . 21
   5.2 Examination of the Armor Data by Panel . . . . . . . . . 25
      5.2.1 Estimation of Individual Panel $V_{50}$s . . . . . . . . . 25
      5.2.2 Bullet Fragmentation Phenomenon . . . . . . . . . 26
   5.3 Effect of Alternative Data Sampling . . . . . . . . . . . . 28
      5.3.1 Effect of Panel B Back . . . . . . . . . . . . . . . . 28
      5.3.2 Effect of Shots with Velocities above 541 m/s . . . 28

6 **Summary and Conclusions**   31

7 **References**   33

This page intentionally left blank.

# List of Tables

2.1 Comparison of distributions.. . . . . . . . . . . . . . . . . . . 8
2.2 Summary of estimates for new UD-PPTA body armor. . . . . 10
3.1 $V_{50}$ estimates and selection criteria for new UD-PPTA armor. 15
4.1 The best models for new and conditioned PBO armors. . . . 20
5.1 Summary of estimates for a new hybrid body armor. . . . . . 23
5.2 Summary of selection criteria for a new hybrid body armor. . 24
5.3 $V_{50}$ estimates for each armor panel. . . . . . . . . . . . . . 26
5.4 Results, excluding Panel B Back. . . . . . . . . . . . . . . . 28
5.5 Results, excluding shots above 541 m/s. . . . . . . . . . . . 29

This page intentionally left blank.

# List of Figures

2.1 Cumulative density functions for the three GLMs. . . . . . . 6
2.2 Probability density functions for the three GLMs. . . . . . . 7
2.3 Estimated response curves for a new UD-PPTA body armor. 9

4.1 Estimated logistic response curves for PBO armors. . . . . . 18
4.2 Estimated probit response curves for PPTA armors. . . . . 19

5.1 Logit estimated response curves for a new hybrid armor. . . . 22
5.2 Probit estimated response curves for a new hybrid armor. . . 22
5.3 C-log-log estimated response curves for a new hybrid armor. . 23
5.4 Estimates of $V_{50}$ using the three different models. . . . . . . . 24
5.5 Probability of perforation at the NIJ reference velocity. . . . 25
5.6 Photograph of de-constructed armor. . . . . . . . . . . . 27
5.7 Photograph of shattered bullet removed from armor. . . . . . 27

This page intentionally left blank.

# 1 Introduction

The performance standard for ballistic-resistant body armor published by the National Institute of Justice (NIJ), NIJ Standard–0101.06[1], recommends estimating the performance of body armor by performing a statistical analysis on $V_{50}$ ballistic limit testing data. During a $V_{50}$ ballistic limit test, bullet velocity is varied up and down to obtain mixed outcomes where some shots are stopped by the armor, but other shots yield perforations of the armor. From these results, the ballistic performance of the armor can be estimated using a statistical regression model. The logistic regression model is commonly used in many binary response systems, such as the investigation of a relationship between age and the presence or absence of a disease for medical applications[2], and is therefore recommended in NIJ Standard-0101.06[1]; but depending on the armor system and the amount of ballistic data collected, other regression models may be more appropriate. Unfortunately, one does not usually have enough ballistic data to determine which model best describes the armor's physical situation. Thus, this study has applied a number of regression models to the data for the purpose of better understanding how well the various models fit the data.

In this work, each regression model will be applied to the ballistic limit data, using the method of maximum likelihood[2,3] to estimate the regression parameters. Then, estimation results from the model will be compared and evaluated. Different criteria for assessing the goodness of each model will be investigated. The objective is to identify criteria that can distinguish which regression method produces the best estimate of the performance of a particular armor model. Another objective of this work is to apply these models to different $V_{50}$ datasets to better understand how an armor model's performance changes as it ages, either through field or laboratory aging. Ballistic limit methods have been commonly applied to provide insight into field-aged armor performance, and many questions still remain regarding whether the initial distribution model, deemed appropriate for new armor, continues to describe the armor as it ages. Results will be presented to explore the selection of an appropriate distribution model for both new and environmentally conditioned armor samples. Three models are considered: the logistic, the probit and the complementary log-log (extreme value type I) response models. The logistic and probit are the two most commonly used methods. The difference between these two symmetrical and asymptotical distributions is known to be small. The extreme value distribution (complementary log-log model), which is asymmetric, produces slightly different estimates than the previous two. Finally, an analysis will be performed on a new armor with a typical $V_{50}$ test results to evaluate and compare the estimations of performance provided by the three different distribution models.

# 2 $V_{50}$ Ballistic Limit Data Analysis

## 2.1 Experimental

For the first part of the study, ballistic limit data was generated in the NIST-OLES (National Institute of Standards and Technology-Law Enforcement Standards Office) ballistic testing laboratory on new armor and environmentally conditioned armor [4]. For the second part of this study, ballistic limit data was generated data a NVLAP (National Voluntary Laboratory Accreditation Program)-accredited ballistic testing laboratory on new armor. During a $V_{50}$ ballistic limit test, the propellant charge weight of each bullet is varied in order to control the bullet velocity. This test procedure is designed to obtain mixed outcomes where the armor stops some shots, but yields perforations on other shots. For each shot, the bullet velocity is recorded and the armor panel is examined to determine whether or not a perforation occurred; the shot outcome is then codified 1 or 0 respectively. All panels were tested according to the procedures detailed in NIJ Standard–0101.06, Section 7 [1]. From these results, the ballistic performance of the armor can be estimated using a statistical regression model and the $V_{50}$ value can be approximated. The $V_{50}$ is the velocity at which a given bullet type has a probability of perforation of 50 %, which is a useful property for comparing different types of armor systems.

The first part of this study focuses on datasets generated from four different body armor models. For each armor model there are test data from both new and environmentally conditioned armor. The four armor types that are considered are body armors containing poly-p-phenylene benzobisoxazole (PBO), p-phenyleneterephtalamide (PPTA), unidirectional p-phenyleneterephtalamide (UD-PPTA) and ultra high molecular weight polyethylene (UHMWPE). These acronyms are used to identify the armors in this work. The second part of the study focuses on datasets generated from a hybrid body armor model which combined PPTA and UHMWPE and had an unusual ballistic limit response as compared to the other armors in the study.

## 2.2 The Logistic, Probit and Complementary Log-Log Regression Models

### 2.2.1 Background

Once the ballistic limit testing has been completed, the test results are analyzed for each threat by performing a regression to estimate the performance of the armor over a range of velocities. During ballistic limit testing only a limited number of shots can practically be taken, and from those data the full performance of the body armor can be estimated. In particular, the analysis attempts to estimate the velocity where the probability of perforation becomes reasonably small. As previously mentioned, the shot outcome of the ballistic limit tests is a perforation or a stop, codified as 1 and 0, respectively. This type of outcome data is commonly called binary response data. A vast literature in statistics, biometrics, and econometrics is concerned with the analysis of binary response data and the classical approach fits a binomial regression model using maximum likelihood[5]. The binomial regression model is a special case of an important family of statistical models, namely Generalized Linear Models[6, 7, 8, 9] (originally due to Nelder and Wedderburn[10]). The acronym GLM is a shorthand for generalized linear model [11]. The binomial family is associated with several links; among the common binomial links there are the probit, the logit and the complementary log-log link functions. The probit model [12] was the first model of binary regression used. It was originally developed for analyzing dose-response data from bioassays[12,13]. This model is still used by researchers for biological assay analysis, and often used to model other data situations. Logistic regression was developed later and not used much until the 1970s, but it is now more popular than the probit model [11,14]. Indeed, in recent decades, the logistic regression model has become the standard method of analysis for binary response data to model the relationship between the binary outcomes (shot outcome in this case) and the independent variable (the velocity of the bullet) in many fields, such as biomedical[2,11,15,16,17] or economical research[14,18]. Alternatives to the traditional logistic approach using the probit and complementary log-log link functions were studied[19,20,21].

Current opinion regarding the selection of link function in binary response models is that the probit and logistic links give essentially similar results[2, 8, 11, 22, 23]. Long[24] wrote that the choice between the logistic and probit models is largely one of convenience and convention, since the substantive results are generally indistinguishable. Moreover, Gill [7] discussed link functions including the complementary log-log and indicated that any of these three link functions can be used and will provide identical substantive conclusions. The close proximity of the probit and logistic link functions is frequently extrapolated to imply that all links are essentially indistinguishable, but this can be a misconception. Conversely, other studies have shown that in many cases this most commonly used logistic regression model may be not always the most appropriate, and that alternative models can also provide good results in this context of binary response data[19,25].

Many authors have examined the best way to discriminate the logit and probit models [24, 25, 26]. Logistic regression is usually preferred because of the wide variety of fit statistics associated to the model. However if normality is involved in the linear relationship, as it often is in bioassay, then probit may be the appropriate model. It may also be used when the researcher is not interested in odds but rather in prediction or classification [8]. Hahn and Soyer [27] found clear evidence that model fit can be improved by the selection of the appropriate link even in small datasets, and that the probit and logit links do not always give similar results in binary data analysis. They showed that in certain cases, the probit model provides a better fit, while in others the logit model is more appropriate. Moreover, empirical support for the recommendations regarding both the similarities and differences between the probit and logit models can be traced back to results obtained by Chambers and Cox [26]. These researchers found that it was only possible to discriminate between the two models when sample sizes were large and certain extreme patterns were observed in the data [26]. Despite the similarities of these models, even minimal differences can lead to different estimations in some particular cases [26]. Thus, it is always recommended to attempt to apply more than one regression model to the data to better understand the abilities of other models to fit those data.

The use of regression models to analyze ballistic tests of armor systems have been suggested in different studies [28, 29]. NIJ Standard–0101.06 [1] recommends the use of the logistic regression model for the analysis of ballistic limit data. However, other probability distributions and regression methods may be used when one can be shown to better estimate the performance of a particular armor model. In the field of analysis of perforation statistics of body armor, Van Es [30] studied the probit method versus the Kneubuhl method and showed that probit analysis was a robust tool to analyze ballistic limit data. Maldague [31] also studied the analysis of $V_{50}$ using different methods and used successfully the probit method with ballistic results. For this reason, other alternative distributions able to fit these data will be studied and compared in terms of quality of estimation of the armor performance.

### 2.2.2 Presentation of the Three GLMs

As previously mentioned, three different distribution models are considered in this study: the logistic, the probit and the complementary log-log (extreme value type I) response models. For the purposes of this paper, the logistic regression model will be called logit or logistic, and the complementary log-log regression model will be called c-log-log. The logit link function is a fairly simple transformation of the prediction curves so it is popular among researchers [22]. Logit models use the logistic probability distribution [22]. The probit models assume the standard normal distribution [12]; it has a mean of 0 and standard deviation of 1. The standard logistic distribution has a mean of 0 and standard deviation of 1.8. When both models fit well, parameter estimates in logistic regression models are approximately 1.8 times those in probit models [24]. The normal and logistic distributions

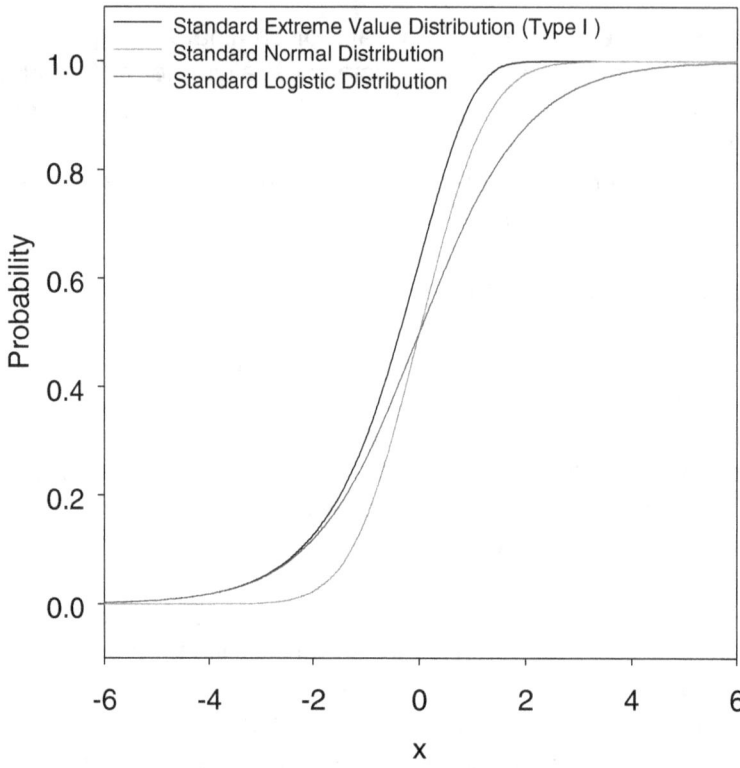

Figure 2.1: Cumulative density functions for the three GLMs.

are both symmetric [12, 22]. The logit and probit links are very similar; in particular, both approach 0 and 1 symmetrically and asymptotically. Because of this similarity, they usually lead to analogous results [12, 22]. The c-log-log analysis is an alternative to logit and probit analysis. The c-log-log model is based on the extreme value type I distribution, also referred to as the Gumbel distribution [32, 33], which is asymmetric in contrast to the logistic or standard normal distribution of the logit and probit models, respectively. All of the three model transforms produce a sigmoidal (or S-shaped) response curve. However, since the extreme value distribution is asymmetric, the results are slightly different from those of the two other symmetrical models. To illustrate this difference, idealized curves for all three models are presented in Figures 2.1 and 2.2. The reason why Cumulative Distribution Functions (CDF) are used as link functions for binary data is because the CDF is always between 0 and 1.

The cumulative density function of the standard normal distribution is steeper in the middle than that of the standard logistic distribution and quickly approaches 0 on the left and 1 on the right. From Figure 2.2 it can be noted that the logit link has heavier tails than the probit or c-log-log, i.e. this link assigns a greater probability to observations which fall outside the mean. The implication of this is that in the event that there is a lot of variability in the measurements the parameter estimates using the logit link will capture this as a result of the heavier tails whereas the other distributions may not. From this observation one could assume that

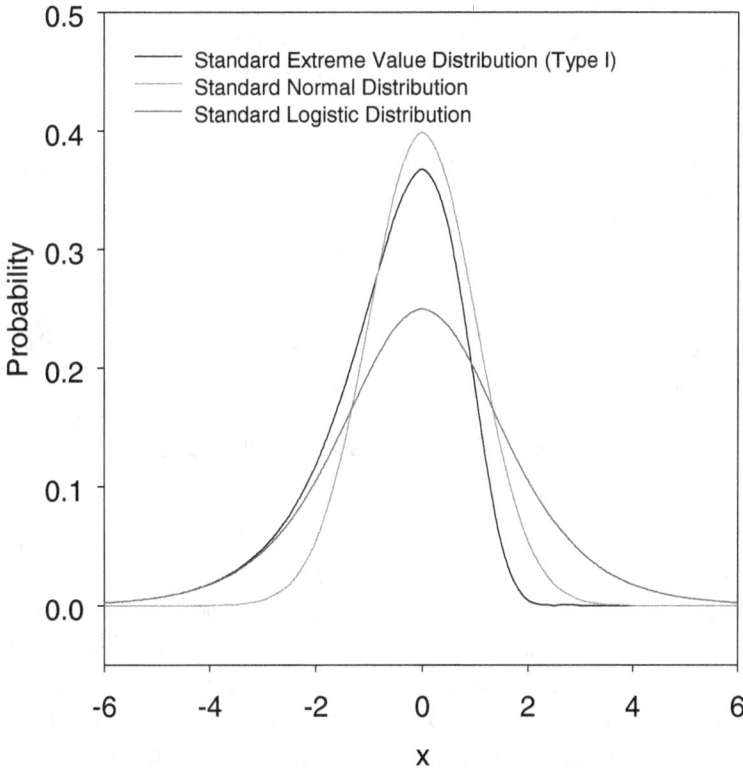

Figure 2.2: Probability density functions for the three GLMs.

it is generally safer to use the logit link as it is less susceptible to outliers or to data with a lot of variability than the two other links. The three link functions and their corresponding distributions are summarized in Table 2.1.

Table 2.1 contains the probability of a complete perforation occurring at velocity $v$: $\pi(v)$; and also the velocity at which the probability of perforation is $\pi\%$: $\hat{V}_\pi$. The calculation of $V_{50}$ is determined from the estimated regression parameters $\beta_0$ and $\beta_1$, which are the estimated *constant* and the estimated *velocity coefficient*, respectively. The formula of this estimated $V_{50}$ is also shown in Table 2.1 for each model. The explanation and the calculation of the confidence intervals of the estimates can be found in Ref. [2, 34]. The different regressions are performed on the data using the method of maximum likelihood [2, 3] to estimate the logistic, the probit or the c-log-log parameters $\beta_0$ and $\beta_1$. The confidence intervals of the estimates are calculated using the Wald test [2, 34]. Fieller's theorem [12] is used to estimate the confidence intervals of $V_{50}$.

## 2.3 Application of the Different Models to Fit $V_{50}$

### 2.3.1 R Software

The three regression models are performed on the ballistic limit data using the method of maximum likelihood to estimate the regression parameters. The R statistical software package [35] was used to execute the different

| Link Name (distribution) | Probability of perforation $\pi(v)$ | Velocity where probability of perforation is $\pi$, $\hat{V}_\pi$ | Estimated $V_{50}$ |
|---|---|---|---|
| Logit (Logistic) | $\dfrac{e^{\beta_0+\beta_1 v}}{1+e^{\beta_0+\beta_1 v}}$ | $\dfrac{\ln\frac{\pi}{1-\pi}-\beta_0}{\beta_1}$ | $\dfrac{-\beta_0}{\beta_1}$ |
| Probit (Normal) | $\Phi(\beta_0+\beta_1 v)$ | $\dfrac{\Phi^{-1}(\pi)-\beta_0}{\beta_1}$ | $\dfrac{-\beta_0}{\beta_1}$ |
| Complementary log-log (Extreme value) | $1-e^{-e^{(\beta_0+\beta_1 v)}}$ | $\dfrac{\ln(-\ln(1-\pi))-\beta_0}{\beta_1}$ | $\dfrac{\ln(-\ln(0.5))-\beta_0}{\beta_1}$ |

Table 2.1: Comparison of distributions.

regressions and estimate the regression coefficients. R is a free software environment for statistical computing. It provides a wide variety of statistical techniques, such as data analysis using regression models. The R statistical software allows computing and fitting each of the three different regression models (logistic, probit and c-log-log) to the data. The generalized linear model (GLM) procedure [36], with the parameters family=binomial and link=logit, probit, or cloglog, as appropriate, to specify the model is used to fit the different regression models to the binary response data using maximum likelihood estimation. After the regression computation, R outputs provide the regression coefficients estimates and their standard errors, as well as all the information needed to calculate the confidence interval (variance matrix), and also some useful statistics like deviance and Akaike's Information Criterion. Both statistics will be discussed further later in this work.

## 2.3.2 Comparison of the GLM Estimates

To illustrate the application of the different regression analysis and their results in terms of estimations, a typical example is presented in Figure 2.3: the analysis of the ballistic performance of a new UD-PPTA body armor which shows the estimated response curve of this body armor given by the three different models.

The estimated response curves of the logistic, probit and c-log-log regression models and their 95% confidence intervals as presented in Figure 2.3 are very similar. However, the confidence interval generated using the c-log-log model is not comparable with those obtained from logistic and probit models because of the asymmetrical shape of the c-log-log distribution function. The three function curves are all S-shaped. As previously discussed, the logit and probit curves are very similar; in particular, both approach 0 and 1 symmetrically and asymptotically. However the c-log-log distribution is asymmetric, it approaches 1 much more rapidly than it approaches 0, accordingly, the results obtained with this model are different. The logistic and probit functions are almost linearly related over the interval of probabilities of perforation between 0.1 and 0.9. These two models perform

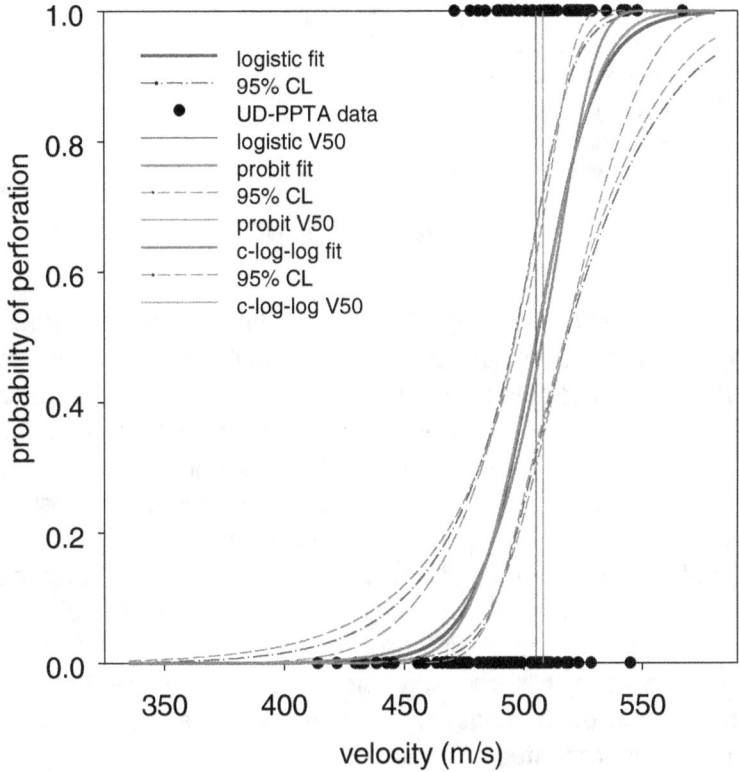

Figure 2.3: Estimated response curves for a new UD-PPTA body armor given by the three GLMs.

|  | Logit | Probit | C-log-log |
|---|---|---|---|
| $V_{50}$ **(m/s)** | **505** | **505** | **508** |
| Upper 95% CL on $V_{50}$ (m/s) | 518 | 517 | 518 |
| Lower 95% CL on $V_{50}$ (m/s) | 496 | 497 | 498 |
| CI width (m/s) | 22 | 21 | 19 |
| **Predicted prob. at 350 m/s** | **0.0000044** | **5.63E-14** | **0.0000653** |
| Upper 95% CL at 350 m/s | 0.002392 | 0.000034 | 0.005925 |
| Lower 95% CL at 350 m/s | 8.01E-09 | 8.51E-28 | 7.18E-07 |
| CI width | 0.002392 | 0.000034 | 0.005924 |
| $V_{02}$ **(m/s)** | **457** | **462** | **448** |
| Upper 95% CL on $V_{02}$ (m/s) | 473 | 477 | 469 |
| Lower 95% CL on $V_{02}$ (m/s) | 406 | 426 | 390 |
| CI width (m/s) | 67 | 50 | 78 |

CI is confidence interval.
CL is confidence level.
350 m/s is the NIJ reference velocity.
$V_{02}$ is the velocity at which a bullet has a 2% chance of perforating the armor.

Table 2.2: Summary of estimates for a new UD-PPTA body armor.

similarly at the center (at $V_{50}$). The primary difference between the logit and probit response curves is that probit has slightly flatter lower and upper asymptotes, which means the probit curve approaches the axes more quickly than the logistic curve. Therefore, the two GLMs give different estimations of armor perforation for low and high bullet velocities, which are important in ballistic limit analysis. As expected, the asymmetrical c-log-log response curve approaches much more quickly the high probability of perforation (i.e., probability of 1) than either the logit or probit function. For small values of probability of perforation, the c-log-log function is close to the logistic. The preceding discussion of the differences and similarities between the response curves given by the three regression models considered here are confirmed by further examining the $V_{50}$ and $V_{02}$ estimate values directly. This information is presented in Table 2.2.

While the estimates of parameters differ in size due to the different scaling of the normal and logistic distributions, the substantive conclusions (and the predicted probabilities of perforation for the armor) are very similar. The estimated $V_{50}$ value is the same for the logit and probit models. Because of its asymmetry, the estimations of the high and low velocities related to the logistic and probit models are different from the ones obtained by the c-log-log model. Generally, the estimated $V_{50}$ provided by the logit and probit regression models are similar but the $V_{50}$ estimated by c-log-log is slightly higher. In the case of the $V_{02}$ estimate, the c-log-log model provides the lowest value for $V_{02}$, while the probit provides the highest value. In most applications, results from the c-log-log model are not very different from logit and probit, however, occasionally the estimate results can suggest qualitatively different conclusions.

Even if differences and similarities between estimates given by the three regression models can be discerned, it is difficult to discriminate between these three models on the basis of the quality of the armor performance estimation. The three regression models lead to very similar results, especially for the estimation of $V_{50}$. For the estimation of $V_{02}$ and the predicted probability of perforation at the NIJ reference velocity (350 m/s), the difference between the estimates given by the diverse models is larger, but they are still similar. Moreover, the binary nature of the analyzed data does not allow a visual comparison. From Figure 2.3 it is not possible to identify the best model. Consequently, some criteria of goodness-of-fit are required to assist in the determination of which model could better estimate the armor performance from the available ballistic data. The next section will examine this issue.

This page intentionally left blank.

# 3 Generalized Linear Model Estimation Evaluation

Different criteria for assessing the goodness-of-fit of each model will be applied to the ballistic limit data. The objective of this analysis is to identify criteria that can distinguish which regression method produces the best estimate of the performance of a particular armor model, since the estimations given by the three models were shown to be analogous.

## 3.1 Assessment Criteria

Once a model has been fitted to the observed values of a binary response variable, it is essential to check that the fitted model is actually valid. Goodness-of-fit statistics given by the R software can be used to compare fits using different link functions. The significance tests of the regression coefficients is provided after every regression and allows one to verify the significance of each coefficient. The R outputs also indicate the estimates of the regression coefficients and are accompanied by the standard errors.

### 3.1.1 Akaike's Information Criterion

One way to choose between different specifications (e.g. between the probit, logit and c-log-log models) is to use a model selection criterion. Akaike[3] defined an *information criterion* commonly known as *Akaike's Information Criterion*, or AIC. This criterion is a measure of goodness-of-fit which takes into account the number of fitted parameters. The formula for calculation of AIC is given in Equation 3.1.

$$AIC = -2\log L + 2p \qquad (3.1)$$

where $\log L$ is the log-likelihood function evaluated of the model parameters and $p$ is the number of model parameters. The AIC is a convenient metric for this analysis because it is given in R's ANOVA (analysis of variance) output. Smaller AIC values are associated with better fits. The AIC was calculated for all the GLMs considered, and the model with the smallest AIC is considered to be the closest to the unknown reality that generated the data.

### 3.1.2 Log-Likelihood

As a quick and simple way to compare the performance across the different models, one can simply look at the maximized log-likelihood of each specification, since the models contain an equal number of parameters. However, Akaike[3] showed that the maximized log-likelihood is biased upward as an estimator of the model selection criterion and then defined the AIC as a better criterion for measuring goodness-of-fit.

### 3.1.3 Deviance of the Model

Huettmann and Linke[21] presented two methods of assessing which link function performs best for inferences and for predictions. The first decision criterion is centered on the model deviance, e.g. relevant for inferences. A measure of discrepancy between the observed and fitted values is the deviance statistic. In a perfect fit the deviance is zero. Thus, the most preferable model can be found on the basis of the minimum-deviance criterion for model selection. For example, if the deviance of a probit model is significantly lower than the one of the corresponding logit model, then the former is preferred. This postulate holds when comparing any of the links within the binomial family[8]. Conversely, the model that provides the least desirable fit to the data can also be found. The deviance criterion is also given in the R output as an indicator of goodness-of-fit.

### 3.1.4 Prediction Error Rate

The second criterion presented by Huettmann and Linke[21] is based on prediction errors. It uses the differences between expected and predicted values as an indication of the fit. Once the regression model analysis is performed, the resulting regression coefficients estimates are used to predict the data and provide predictive probabilities of perforation. If the probabilities of perforation are greater than 50%, these probabilities are classified in the perforation group, if they are less than 50%, they are classified in the non-perforation group. Then the observed and predicted responses can be cross-tabulated and the proportion of cases predicted correctly can be calculated. The lower the misclassification rate, the better the model fits the data. However, this misclassification rate is not independent of the model (since it is based on the data used to build the model) and therefore could underestimate the real error rate.

### 3.1.5 Cross-Validation Method

The cross-validation method avoids this problem of dependence; therefore, it gives a better calculated error rate than the usual prediction error rate. In $k$-fold cross-validation, one divides the data into $k$ subsets of (approximately) equal size. One trains the net $k$ times, each time leaving out one of the subsets from training, but using only the omitted subset to compute the error criterion of interest, in this case the prediction error rate. This way, it

|  | Logit | Probit | C-log-log |
|---|---|---|---|
| $V_{50}$ (m/s) | **505** | **505** | **508** |
| Upper 95% CL on $V_{50}$ (m/s) | 518 | 517 | 518 |
| Lower 95% CL on $V_{50}$ (m/s) | 496 | 497 | 498 |
| AIC | 51.94 | **51.40** | 52.07 |
| Misclassified Data (%) | 23.3 | 23.3 | **21.7** |
| Error Rate by Cross-Validation (%) | 23.3 | 23.3 | **21.7** |

Table 3.1: $V_{50}$ estimates and selection criteria for new UD-PPTA armor.

avoids the problem of dependence of the model observed previously. Cross-validation can be used simply for model selection by choosing the model that has the smallest estimated generalization error.

In this study, where the ballistic performance of an armor type is estimated by testing several armor panels and combining their ballistic results, this training/testing set method uses the ballistic data of $n-1$ panels to estimate the model and uses the data of the $n^{th}$ panel to test the prediction of the model. Then the model is trained $n$ times, each time leaving out one panel to calculate the error rate. Finally the average error rate of re-substitution is calculated and used as a model selection criterion.

Among all of the model selection criteria previously presented, the AIC and the average error rate calculated by cross-validation, also called average error rate of training/testing sets method, will be used as criteria to try to distinguish which regression method produces the best estimate of the performance of a particular armor model. However, in case of few differences between the model results, the optimal model could be at the end chosen by the user, regarding mostly its specifications, its estimations and its applications in a practical way.

## 3.2 Model Diagnostics Results

Some general observations about the results of these model diagnostics can be noted. First, the lack of fit tests given by the R software output do not indicate a significant lack of fit for any of the three models, for all the data sets tested, and the criteria values for all three models are similar. Furthermore, the prediction error rate is, as expected, lower when it is calculated from the whole set of data used to create the model, and the average error rate of cross-validation method is higher. These misclassification rate criteria have the same values for logit and probit. The comparison of the predicted values, previously made in Figure 2.3 showing the predicted values given by the three models and the observed data, did not indicate strong evidence for distinguishing models on the basis of link. Therefore, the model selection criteria given in Table 3.1 will be the focus of the rest of this work.

In Table 3.1, the best criteria are shown in bold. Thus the AIC statistic for the probit model is lower than for the logistic and c-log-log models, suggesting a slight, but probably insignificant, preference for the probit model.

The criteria based on the two error rates show that the c-log-log model is better for prediction. However the difference between the error rates for the different models is small.

## 3.3 Results of Theoretical Analysis

A regression model can be fitted to the ballistic limit data to estimate the overall response of particular body armor. Different possible models were considered and compared in terms of quality of estimation of the body armor's performance. Comparisons were made only between models that have been applied to the same dataset, so with the same number of data points and the same number of parameters. The comparison was made at the level of estimation of the $V_{50}$. The estimation given by each model was evaluated on its confidence interval (CI) width, the percentage of misclassified data, the estimation of $V_{50}$ and other parameters like AIC. The results show that the logistic analysis generally gives a good overall estimation of the body armor's performance, but other regression models could be better. For a particular armor, a model can be shown to better estimate its performance on the basis of a selection criteria. However, the difference between the three models examined here in is relatively small in terms of estimation. Additionally, the values of the different criteria studied are too close to make a meaningful determination among the various models. Even if some criteria indicate a slight preference for one model, the others cannot necessarily be considered bad models to fit the ballistic data. Thus, the commonly used logistic regression model specified in NIJ Standard–0101.06 can still be considered an appropriate choice for $V_{50}$ ballistic limit analysis.

# 4 Application to New and Conditioned Armor Analysis

Within the body armor community, many questions still remain regarding whether the initial distribution model deemed appropriate for new armor continues to describe the armor as it ages. Therefore, the analysis of new and environmentally conditioned armors is examined, with the objective of selecting an appropriate distribution model for both new and environmentally conditioned armor samples. Further details about the environmental conditioning of the body armor discussed herein can be found in Forster and al [4]. Using the three different regression models and their specificities in terms of data fitting presented previously, and the criteria able to help identify the best model for a particular armor, the focus is made on the aspect of body armor conditioning and the fitting of ballistic data from new and environmentally conditioned body armor. Data generated from the $V_{50}$ ballistic limit testing of the new and conditioned samples of the same model of armor are considered.

To facilitate an understanding of the effects of conditioning on the ballistic performance of a particular armor, the estimates provided by the same regression model on new and environmentally conditioned armor data are compared.

## 4.1 Comparison of New and Conditioned Armors

If the results from the new armors are compared to the environmentally conditioned armors, the armor model's performance appears to decrease after conditioning. Whatever the regression model applied, the estimates for $V_{50}$ and $V_{02}$ decrease, while the size of their associated confidence intervals increase. This indicates a reduced confidence in the armor response curve and therefore an increase in the probability of perforation for these velocities. The probability of perforation at the NIJ reference velocity also increases, as well as its confidence interval. The observed shifts in values for $V_{50}$ and $V_{02}$ are of the same range, regardless of which regression model is used. To illustrate these observations, the observed data and the predicted probability of perforation given by the logistic regression for new and environmentally conditioned PBO armors are presented in Figure 4.1. The new armor is

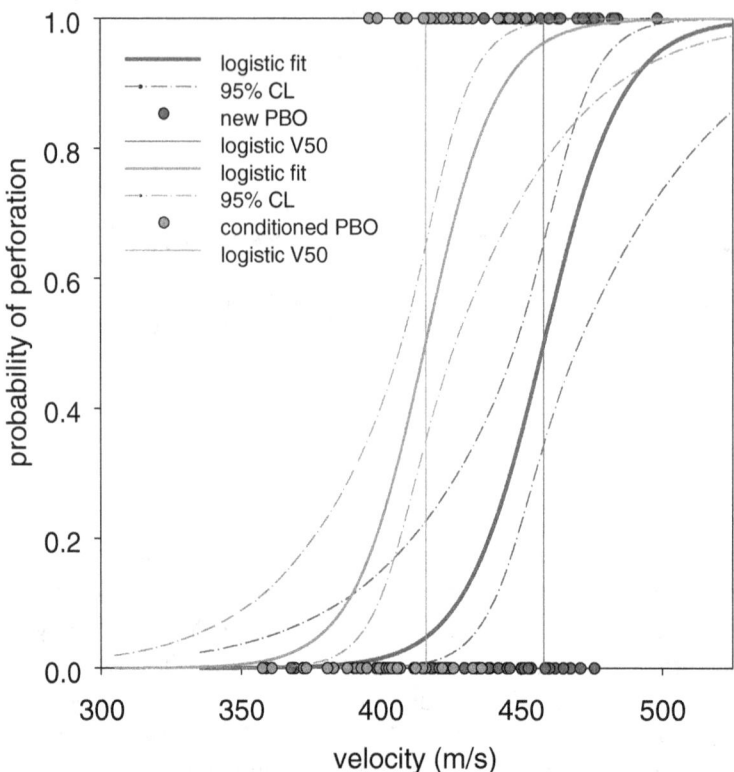

Figure 4.1: Estimated logistic response curves for new and conditioned PBO armors.

represented in blue and the environmentally conditioned armor in red. The shape of the ballistic response curves of new and environmentally conditioned armors looks similar, but as previously mentioned, it appears that the response curve has shifted to the left when the armor is environmentally conditioned.

Obviously the PPTA armor is made from a different material than the PBO armor, subsequently the ballistic performances of new and environmentally conditioned PPTA armors are different. The point estimates for $V_{50}$ of conditioned PPTA armor are higher than those for new PPTA armor, though the size of the linked confidence intervals increases as well. However, the $V_{02}$ decreases and the probability of perforation at NIJ reference velocity increases in the PPTA armor, as was seen in the PBO armor, and the confidence intervals associated with the analysis also increase. Figure 4.1 illustrates these observations. Note the shape of estimated response curve changes between the new and the environmentally conditioned armors. The curve of environmentally conditioned armor is more elongated and its slope is less steep. The confidence interval of the environmentally conditioned armor response curve is much wider, so the uncertainty in the ballistic performance increases as the PPTA armor is conditioned.

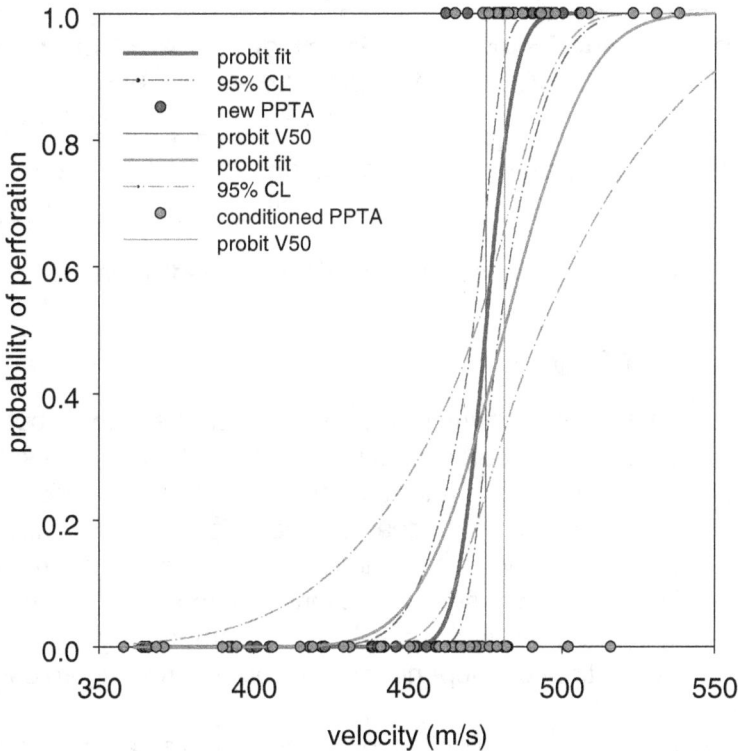

Figure 4.2: Estimated probit response curves for new and conditioned PPTA armors.

## 4.2 Selection of an Appropriate Model for New and Environmentally Conditioned Body Armors

This study seeks to answer the question of whether or not the initial distribution model deemed appropriate for new armor continues to describe the armor as it ages, and the goal is to try to find an appropriate distribution model for both new and environmentally conditioned armor samples. Typical results of the selection model criteria for PBO new and environmentally conditioned armors are shown in Table 4.1. Using the prediction performance as a criterion, the three models behave similarly. Using the model AIC as a decision criterion, the findings indicate that for the $V_{50}$ ballistic data studied, the probit model would best fit the data. On the basis of the minimum AIC and the minimum average error rate criteria for model selection, the probit model could be slightly preferable to the others and is deemed appropriate for both new and environmentally conditioned armors. However, as previously noted, the difference between the models is minimal, especially between the probit and logistic models.

| Armor Type | Minimum AIC | Minimum Misclassification (%) | Minimum average Error Rate by Cross-Validation |
|---|---|---|---|
| New | probit | c-log-log | tie |
| Conditioned | probit | logit/probit | tie |

Table 4.1: The best models for new and conditioned PBO armors.

## 4.3 Summary

The results show that for a particular armor, one GLM can be shown to better estimate the performance of that specific armor. However, the difference between the GLMs is relatively small in terms of the quality of estimation, and the values of the different criteria studied are similar. Thus, even if some criteria prefer one GLM, the other GLMs could still be an appropriate choice to fit the ballistic dataset. None of the three models examined herein can be generally considered to be the *best* model - that is, the model providing the best estimate of the performance of body armor, regardless of its condition.

The contingency table (also called confusion matrix) is a table presenting observed perforations and stops versus their predicted values (in this case predicted perforations or stops). In this matrix, the number of correctly and incorrectly predicted data points is presented. As far as the ballistic limit analysis is concerned, the amount of misclassified perforations is an important number because it represents the amount of real perforations that are not estimated well by the model. Considering that body armor is life safety equipment, one could assume the viewpoint that it is more serious to misclassify an observed perforation as a predicted stop than the opposite.

By examining the contingency table for all the different models applied to the ballistic limit data, it can be noted that the estimates of logit and probit models have the same amount of misclassified perforations, and that the c-log-log model estimates have the same or slightly more misclassified perforations. Consequently, either the logit or the probit models may be slightly preferred over the c-log-log model, though the c-log-log model cannot be considered to be invalid based on this observation.

# 5 Application of the GLMs to an Unusual Data Set

In the previous sections, the analysis with the different regression models did not show much difference between the models. However, if an armor had an atypical ballistic limit response, then perhaps one of the GLMs could be shown to better describe that particular armor. To investigate this possibility, a data set from an armor in which one panel had a high number of stops or perforations was of interest in determining if any of the three different models could be determined to be more appropriate than the others for such a system.

## 5.1 Global Analysis of the Armor

Data from a new hybrid armor model with a large number of high-velocity stops on one panel were selected for this analysis. The estimated response curves for all three regression models are presented in Figures 5.1, 5.2, and 5.3. As in the previous analysis, the estimated response curves of the logistic, probit and c-log-log regression models and their 95% confidence intervals presented in Figures 5.1 through 5.3 are very similar. However, the confidence interval generated using the c-log-log model is not comparable with those obtained from logistic and probit models because of the asymmetrical shape of the c-log-log distribution function. As previously discussed, the logit and probit curves are very similar; in particular, both approach 0 and 1 symmetrically and asymptotically. However the c-log-log is asymmetric, meaning that it approaches 1 much more rapidly than it approaches 0.

It is important to note that this particular armor model has a very large zone of mixed results (ZMR) of 76.2 m/s (250 ft/s). This is due to a large number of stops on one particular armor panel at high velocities. This large zone of mixed results makes it difficult for any model to accurately predict the armor performance at low velocities by reducing the slope of the curve. These results are presented in Table 5.1.

While the estimates of parameters differ in size due to different scaling of the normal and logistic distributions, the estimated $V_{50}$ values predicted by all three models are very similar (Figure 5.4). The calculation of the error

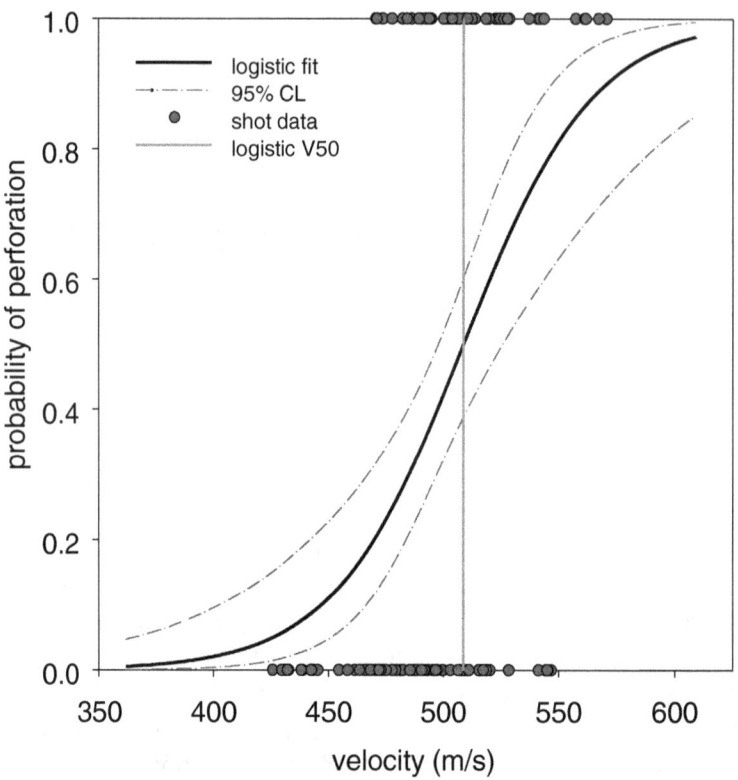

Figure 5.1: Logit estimated response curves for a new hybrid armor.

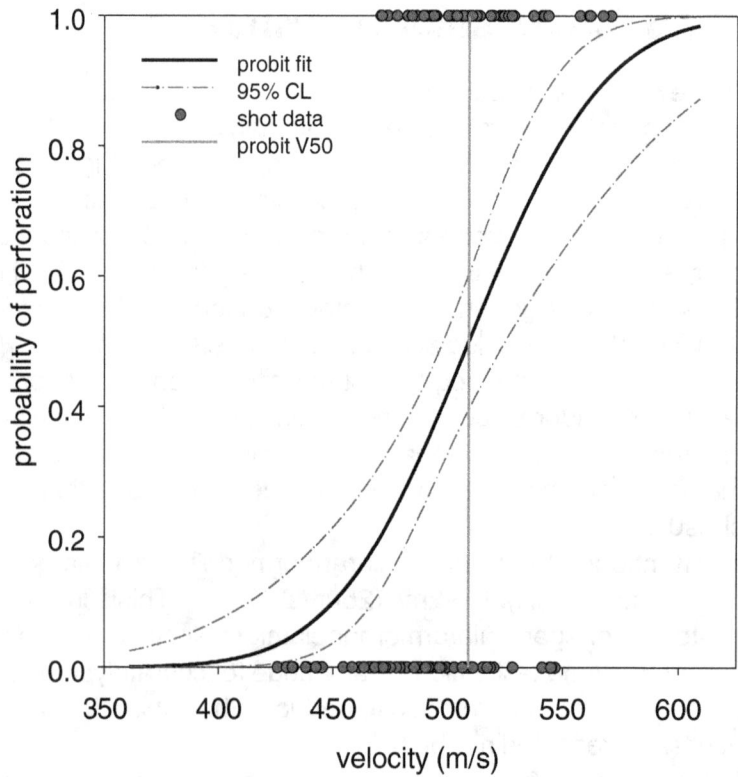

Figure 5.2: Probit estimated response curves for a new hybrid armor.

# 5. Application of the GLMs to an Unusual Data Set

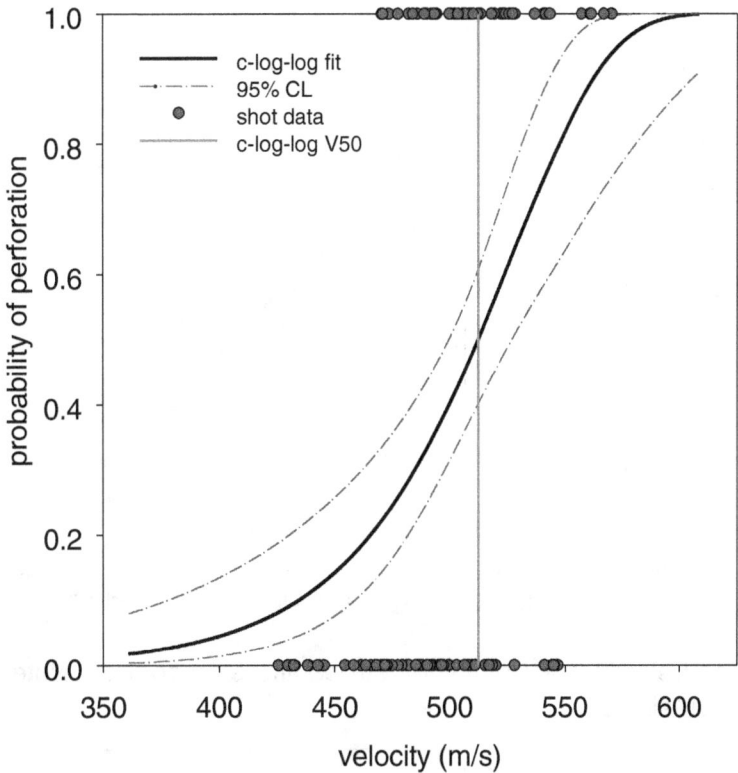

Figure 5.3: C-log-log estimated response curves for a new hybrid armor.

|  | Logit | Probit | C-log-log |
|---|---|---|---|
| $V_{50}$ **(m/s)** | **509** | **509** | **513** |
| Upper 95% CL on $V_{50}$ (m/s) | 525 | 525 | 527 |
| Lower 95% CL on $V_{50}$ (m/s) | 497 | 498 | 500 |
| CI width (m/s) | 28 | 27 | 28 |
| **Predicted probability at 436 m/s** | **0.069745** | **0.055381** | **0.103460** |
| Upper 95% CL at 436 m/s | 0.181378 | 0.163059 | 0.214941 |
| Lower 95% CL at 436 m/s | 0.024743 | 0.013636 | 0.048092 |
| CI width | 0.156636 | 0.149423 | 0.166848 |
| $V_{05}$ **(m/s)** | **426** | **434** | **405** |
| Upper 95% CL on $V_{05}$ (m/s) | 450 | 455 | 437 |
| Lower 95% CL on $V_{05}$ (m/s) | 365 | 386 | 326 |
| CI width (m/s) | 86 | 69 | 111 |

Table 5.1: Summary of estimates for a new hybrid body armor.

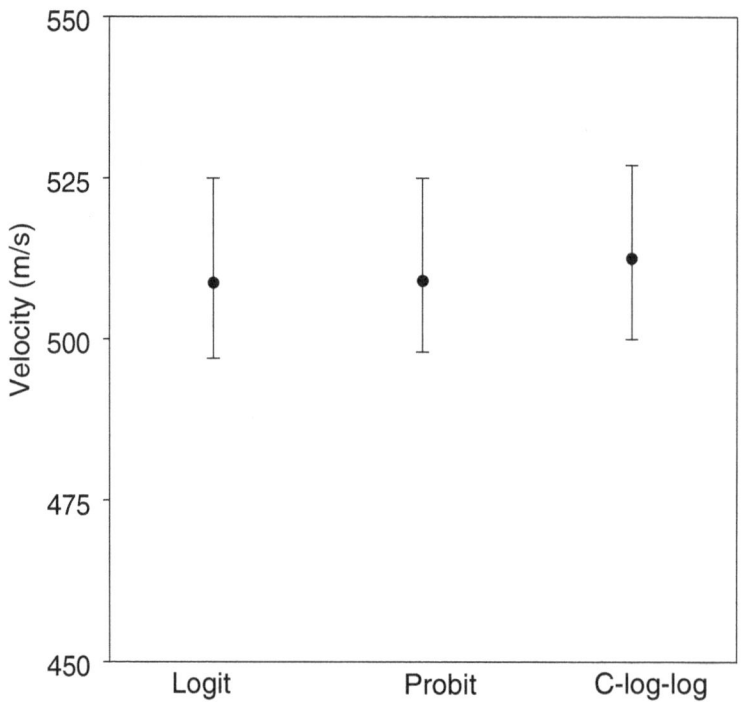

Figure 5.4: Estimates of $V_{50}$ using the three different models. Note that the estimates are similar.

|  | Logit | Probit | C-log-log |
|---|---|---|---|
| **AIC** | 139.61 | **139.15** | 141.06 |
| **Misclassified Data (%)** | **30.0** | 30.8 | 32.5 |
| **Error Rate by Cross-Validation (%)** | 31.7 | 31.7 | 31.7 |

Table 5.2: Summary of selection criteria for a new hybrid body armor.

bars of the $V_{50}$ estimates are based on the Fieller's theorem [12].

However, the probabilities of perforation at the NIJ reference velocity (436 m/s) predicted by the three different models are very different (Figure 5.5). Note that the estimates shown in Figure 5.5 are different and that the error bars (indicating the range of the estimate), calculated using the Wald test [2, 34], are large. This means that, when the error of the estimate is taken into account, all three estimates are probably within the same range. This phenomenon is attributed to the large ZMR for this sample.

Analysis with the c-log-log asymmetric model indicated that the probability of perforation at the NIJ reference velocity was approximately 10% for this armor. This is higher than the 7% that was predicted by the logistical model used in the NIJ Standard–0101.06 ballistic limit calculation, and much higher than the 5.5% that was predicted by the probit model.

Table 5.2 shows the values of the different goodness-of-fit criteria for each model. As in the previous analysis, an assessment of how well the different link functions fit the data using the AIC does not indicate one GLM is a better fit than the others. While there is a slight preference for

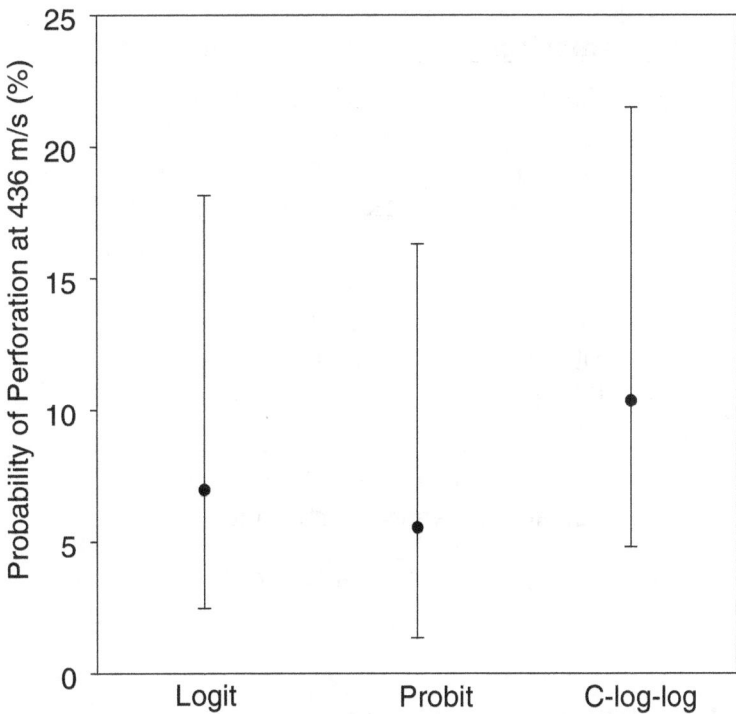

Figure 5.5: Estimates of probability of perforation at the NIJ reference velocity (436 m/s).

the probit model, the difference is not so large as to be significant. This same conclusion can be drawn from an analysis of the misclassification percentage or the average error rate as statistical measures for comparison. The values of these criteria are similar for all three GLMs, however, the c-log-log model has the worst criteria values, so perhaps either the logit or the probit models would be slightly preferred.

## 5.2 Examination of the Armor Data by Panel

### 5.2.1 Estimation of Individual Panel $V_{50}$s

In an effort to better understand the wide ZMR of this armor model, especially the large number of stops on one panel, the individual $V_{50}$ of each armor panel was computed. The estimates were determined using the logistic model. It is important to note that due to the small number of shots on each armor panel, these individual $V_{50}$ estimates are uncertain, but still useful for comparison purposes and for the detection of any anomalies in the armor testing.

From Table 5.3, one can note that Panel B Back had a much higher $V_{50}$ than any of the other panels tested. Additional analysis will further examine this observation.

| ArmorPanel | $V_{50}$ (ft/s) | $V_{50}$ (m/s) |
|---|---|---|
| AFront * | 1583.56 | 482.7 |
| ABack * | 1584.04 | 482.8 |
| BFront * | 1577.24 | 480.7 |
| **BBack ** | **1810.75** | **551.9** |
| CFront ** | 1689.61 | 515.0 |
| CBack ** | 1689.21 | 514.9 |
| DFront ** | 1614.10 | 492.0 |
| DBack ** | 1708.64 | 520.8 |
| EFront ** | 1646.14 | 501.7 |
| EBack ** | 1681.76 | 512.6 |

\* indicates test was conducted on Day 1.
\*\* indicates test was performed on Day 2.

Table 5.3: Logistic $V_{50}$ estimates for each armor panel.

## 5.2.2 Bullet Fragmentation Phenomenon

Another possible explanation for the high $V_{50}$ for Panel BBack may be the unique combination of bullet and target leading to unusual bullet behavior. When this phenomenon occurs, the bullet behavior changes as a function of velocity. For example, at lower velocities, the bullet may deform in a predictable manner as we typically see in armor testing; while at higher velocities, the bullet may fragment upon impact before significant deformation, or perhaps components of the bullet may not remain intact[37, 38]. Because the bullet behavior is different, its penetrative characteristics may be different. This behavior is more typically encountered when testing hard plate armors, but may be possible to encounter when testing at high bullet velocities. Due to the two different competing mechanisms that dominate penetration mechanics, bullet fragmentation effectively leads to more than one ballistic penetration curve for a particular armor-bullet system. In one case, bullet properties dominate (when the bullet breaks up instead of deforming); in another, armor properties dominate (typical armor testing). The armor community typically focuses on finding the lowest $V_{50}$ of the armor system, since that is the one of most practical importance to the person wearing the armor, which is what the test methods in NIJ Standard–0101.06 are intended to do. Note that very high velocities resulted in *stops*, leading one to suspect that the bullets may either fracturing or losing their copper jackets at these very high velocities. To examine this possibility, the armor panel was de-constructed, and bullets were recovered from the armor and examined. These bullets (Figures 5.6 and 5.7) show evidence of bullet failure due to bullet fragmentation. Therefore, additional analysis was performed to examine the effect of this panel on the large ZMR and the outcome of this analysis is the subject of the next section.

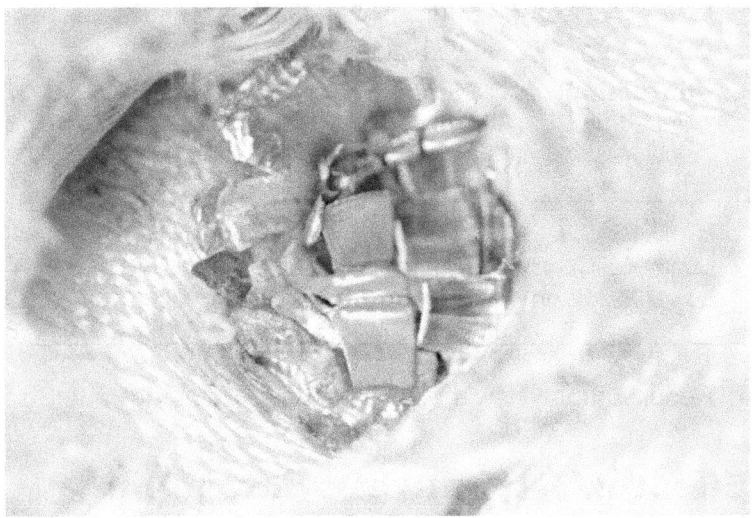

Figure 5.6: Photograph of de-constructed armor with shattered bullet inside.

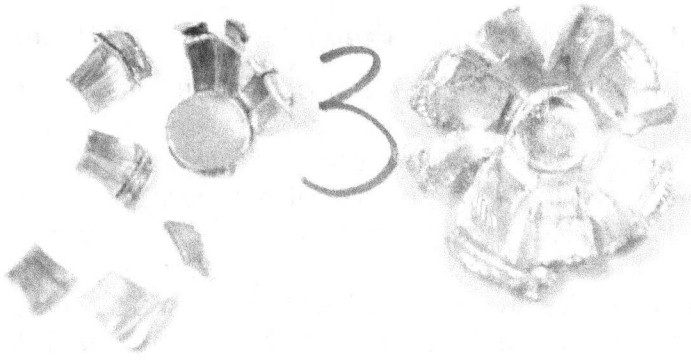

Figure 5.7: Photograph of shattered bullet removed from armor.

|  | Logit | Probit | C-log-log |
|---|---|---|---|
| $V_{50}$ (m/s) | **502** | **502** | **505** |
| Upper 95% CL on $V_{50}$ (m/s) | 514 | 514 | 516 |
| Lower 95% CL on $V_{50}$ (m/s) | 492 | 493 | 494 |
| CI width (m/s) | 22 | 21 | 22 |
| **Predicted probability at 436 m/s** | **0.042333** | **0.029177** | **0.073777** |
| Upper 95% CL at 436 m/s | 0.143464 | 0.124872 | 0.182044 |
| Lower 95% CL at 436 m/s | 0.011532 | 0.004206 | 0.028807 |
| CI width | 0.131932 | 0.120666 | 0.153237 |
| $V_{05}$ (m/s) | **440** | **445** | **423** |
| Upper 95% CL on $V_{05}$ (m/s) | 458 | 461 | 448 |
| Lower 95% CL on $V_{05}$ (m/s) | 393 | 408 | 363 |
| CI width (m/s) | 65 | 54 | 85 |

Table 5.4: Results of analysis with all three models, excluding Panel B Back.

## 5.3 Effect of Alternative Data Sampling

### 5.3.1 Effect of Panel B Back

Since it was confirmed that Panel B Back exhibited anomalous behavior during the test series, the analysis was repeated, excluding the data from Panel B to determine the effect of this panel on the outcome of the analysis. These data are summarized in Table 5.4.

Removing this panel does affect the outcome of this analysis. The logit predicts a 4.23% probability of perforation at the NIJ reference velocity and the probit predicts a 2.92% probability of perforation. Both of these results would meet the criteria required by the NIJ standard. The c-log-log model predicts a 7.38% probability of perforation, but there is no information to indicate that this armor model is best described by an asymmetric distribution.

### 5.3.2 Effect of Shots with Velocities above 541 m/s

All shots above 541 m/s (1775 ft/s) were arbitrarily excluded (resulting in the exclusion of 13 data points) and the analysis was repeated (Table 5.5). The examination of the armor, which indicated that the bullet behavior may have changed dramatically as a function of velocity (due to bullet fragmentation) can be used to justify this approach. Again, the logistic regression analysis indicated that the probability of perforation at the NIJ reference velocity is less than 5%. The logit predicts a 3.58% probability of perforation at the NIJ reference velocity and the probit predicts a 2.22% probability of perforation. The c-log-log model predicts a 5.39% probability of perforation, but again, there is no information to indicate that this armor model is best described by an asymmetric distribution.

|  | Logit | Probit | C-log-log |
|---|---|---|---|
| $V_{50}$ **(m/s)** | **502** | **502** | **504** |
| Upper 95% CL on $V_{50}$ (m/s) | 515 | 515 | 515 |
| Lower 95% CL on $V_{50}$ (m/s) | 493 | 493 | 495 |
| CI width (m/s) | 22 | 21 | 20 |
| **Predicted probability at 436 m/s** | **0.035811** | **0.022165** | **0.053903** |
| Upper 95% CL at 436 m/s | 0.132794 | 0.111885 | 0.152016 |
| Lower 95% CL at 436 m/s | 0.008928 | 0.002513 | 0.018448 |
| CI width | 0.123867 | 0.109372 | 0.133568 |
| $V_{05}$ **(m/s)** | **443** | **448** | **434** |
| Upper 95% CL on $V_{05}$ (m/s) | 461 | 463 | 455 |
| Lower 95% CL on $V_{05}$ (m/s) | 397 | 412 | 381 |
| CI width (m/s) | 64 | 51 | 74 |

Table 5.5: Results of analysis with all three models, excluding shots above 541 m/s.

This page intentionally left blank.

# 6 Summary and Conclusions

This work shows that the choice of link function, between the logit, the probit and the complementary log-log link functions, is not the most important issue in $V_{50}$ ballistic limit performance estimation, since the different GLMs examined all gave similar results. The three regression models have been applied to the ballistic data and then evaluated, but none of them distinguished itself from the others in terms of armor performance estimation. Findings indicate that for all the ballistic data studied, all three link functions behave similarly, even if the model selection criteria prefer a particular regression model. The diverse criteria calculated for all three models were of the same magnitude; therefore, even if one model has a lower value for a criterion, the two other models' criterion values are close. Overall, it can be concluded that for the ballistic data sets examined herein, the logit and probit link functions performed well and seemed to give more accurate estimation of the ballistic performances than the c-log-log function.

The primary objective of this study was to analyze the three regression models to determine which model produces a good estimate of the performance of a body armor model, and to understand how an armor model's performance changes with environmental conditioning. Slight preference can be assigned to the probit and logistic models for new armor, because they gave consistently good results. The comparison of $V_{50}$ ballistic performance results of new and environmentally conditioned armors shows that in general the armor's model performance decreases as it is conditioned. Moreover, all three regression models are appropriate distribution models for both new and aged armor samples. If an initial distribution model is deemed appropriate for new armor, it will continue to describe well the armor as it ages.

The second objective of this study was to examine the usefulness of applying different models to ballistic limit data analysis of a new armor with ballistic limit test results that may indicate that the logistic model is not the appropriate model for this armor. The detailed analysis of $V_{50}$ data from a new hybrid armor, to examine the effect of symmetric and asymmetric regression models (logit, probit, c-log-log) on the predicted performance of the armor at the NIJ reference velocity, showed no effect on the test outcome. Furthermore, there is no evidence to indicate that this armor model is better described by an asymmetric regression model than a symmetric one. However, in the course of completing this analysis, one panel, Armor Panel B back, appeared to have a high number of high velocity stops. Possible explanations of this observation were discussed, including a bullet fragmentation phenomenon, or test anomalies occurring during the test in the laboratory. The high velocity stops observed on this panel contribute to a wide ZMR in the ballistic limit calculations, causing the probability of perforation at the NIJ reference velocity to be higher than the acceptable (5%) criteria. In an effort to better understand the effect of this panel on the test outcome, analysis was repeated with all three models in two different ways: excluding Panel B back from the calculation, and excluding shots above 541 m/s from the calculation. However, the exclusion of either data set is justifiable only if it can be shown that the tests or test conditions were different from what is specified in the NIJ standard.

# 7 References

[1] National Institute of Justice. *Ballistic Resistance of Body Armor*. NIJ Standard–0101.06. U.S. Department of Justice, Office of Justice Programs, Washington, DC, July 2008.

[2] D. W. Hosmer and S. Lemeshow. *Applied Logistic Regression*. John Wiley & Sons, NY, 2nd edition, 2000.

[3] H. Akaike. Information theory and an extension of the maximum likelihood principle. In *Proceedings of the Second International Symposium on Information Theory*, pages 610–624, New York, 1973.

[4] A. L. Forster, K. D. Rice, M. A. Riley, G. Messin, S. Petit, C. Clerici, G. Holmes, and J. W. Chin. Development of soft armor conditioning protocols for NIJ–0101.06: Analytical results. *NISTIR 7627*, 2009.

[5] J. S. Albert and S. Chib. Bayesian analysis of binary and polychotomous response data. *Journal of the American Statistical Association*, 88(422):669–679, 1993.

[6] A. J. Dobson. *An Introduction to Generalized Linear Models*. Chapman & Hall, Boca Raton, FL, 2nd edition, 2001.

[7] J. Gill. Generalized linear models: a unified approach. In *Sage University Papers Series Volume 134*, Thousand Oaks, CA, 2001.

[8] J. W. Hardin and J. M. Hilbe. *Generalized Linear Models and Extensions*. Stat Press Publication, College Station, TX, 2nd edition, 2007.

[9] P. McCullagh and J. A. Nelder. *Generalized Linear Models*. Chapman & Hall, London, 2nd edition, 1989.

[10] J. A. Nelder and R. W. M. Wedderburn. Generalized linear models. *Journal of The Royal Statistical Society. Series A (General)*, 135(3):370–384, 1972.

[11] A. Agresti. *Introduction to Categorical Data Analysis*. John Wiley & Sons, 2nd edition, 2007.

[12] D. J. Finney. *Probit Analysis*. Cambridge University Press, Cambridge, UK, 3rd edition, 1971.

[13] B.G.Greenberg. Chester I. Bliss, 1899-1979. *International Statistical Review*, 8(1):135–136, 1980.

[14] J.S. Cramer. *Logit Models from Economics and Other Fields, Chapter 9: Origin and Development of the Probit and Logit Models*. Cambridge University Press, Cambridge, UK, 1971.

[15] D. G. Altman. *Practical Statistics for Medical Research*. Chapman & Hall, 1991.

[16] P. Armitage, G. Berry, and J.N.S. Matthews. *Statistical Methods in Medical Research*. Blackwell, Oxford, 4th edition, 2002.

[17] F. E. Jr Harrell. *Regression Modeling Strategies: with Applications to Linear Models, Logistic Regression, and Survival Analysis*. Springer, 2001.

[18] J.L. Horowitz and N.E. Savin. Binary response models: Logits, probits and semiparametrics. *Journal of Economic Perspectives, American Economic Association*, 15(4):43–56, Fall 2001.

[19] A. Agresti. *Categorical Data Analysis, chapter 6: Building and Applying Logistic Regression Models*. John Wiley & Sons, Hoboken, NJ, 2nd edition, 2002.

[20] D. Collett. *Modelling Binary Data*. Chapman & Hall, Boca Raton, FL, 2nd edition, 2003.

[21] F. Huettmann and J. Linke. *Assessment of Different Link Functions for Modeling Binary Data to Derive Sound Inferences and Predictions*. Springer, Berlin, 2003.

[22] D.R. Cox and E.J. Snell. *Analysis of Binary Data*. Chapman & Hall, NY, 2nd edition, 1989.

[23] W. H. Greene. *Econometric Analysis*. Prentice Hall, Upper Saddle River, NJ, 3rd edition, 1997.

[24] S. J. Long. *Regression Models for Categorical and Limited Dependent Variables (Advanced Quantitative Techniques in the Social Sciences)*. Sage Publications, Thousand Oaks, CA, 1997.

[25] A.C. Cameron and P.K. Trivedi. *Microeconomics Using Stata*. Stata Press, College Station, TX, 2009.

[26] E. A. Chambers and D. R. Cox. Discrimination between alternative binary response models. *Biometrika*, 54:573–578, 1967.

[27] E.D. Hahn and R. Soyer. Probit and logit models: Differences in a multivariate realm. *The Journal of the Royal Statistical Society, Series B*, 2005. URL http://home.gwu.edu/~soyer/mv1h.pdf. Working Paper.

# REFERENCES

[28] B. P. Kneubühl. Improved test procedures for body armour. In *Proceedings of the Personal Armour Systems Symposium*, Colchester, UK, 1996.

[29] P. L. Gotts, P. M. Fenne, and D. W. Leeming. The application of critical performance analysis to UK military and police personal armour. In *Proceedings of the Personal Armour Systems Symposium*, The Hague, Netherlands, 2004.

[30] M. Van Es. Improved method to determine the stop velocity of armour. In *Proceedings of the Personal Armour Systems Symposium*, Leeds, UK, 2006.

[31] M. Maldague. Evaluation of some methods in order to determine v50. In *Proceedings of the Personal Armour Systems Symposium*, Brussels, Belgiums, 2008.

[32] E. J. Gumbel. *Statistical Theory of Extreme Values and Some Practical Applications*. National Bureau of Standards Applied Mathematics Series 33, Washington, DC, 1954.

[33] NIST/SEMATECH e-Handbook of Statistical Methods, 2010. URL http://www.itl.nist.gov/div898/handbook/ .

[34] D. Leber. Assessment of body armor characteristics. National Institute of Standards and Technology internal communication, March 2006.

[35] R Development Core Team. *R: A Language and Environment for Statistical Computing*. R Foundation for Statistical Computing, Vienna, Austria, 2009. URL http://www.R-project.org .

[36] W. N. Venables and B. D. Ripley. *Modern Applied Statistics with S*. Springer, 4th edition, 2002.

[37] Department of Defense. *Test Method Standard for Performance Requirements and Testing of Body Armor*. MIL-STD-3027. U.S. Department of Defense, Washington, DC, January 2007.

[38] NATO Standardization Agency, Draft STANAG. *Procedures for Evaluating the Protection Levels of Logistic and Light Armoured Vehicles*. AEP-55. NATO/PFP Unclassified, 1st edition, January 2004.